Learn

# Eureka Math™
## Grade 4
## Module 4

**Published by Great Minds®.**

Copyright © 2018 Great Minds®.

Printed in the U.S.A.
This book may be purchased from the publisher at eureka-math.org.
10  9  8  7  6  5  4

ISBN 978-1-64054-067-5

G4-M4-L-05.2018

# Learn ◆ Practice ◆ Succeed

*Eureka Math*™ student materials for *A Story of Units*® (K–5) are available in the *Learn, Practice, Succeed* trio. This series supports differentiation and remediation while keeping student materials organized and accessible. Educators will find that the *Learn, Practice,* and *Succeed* series also offers coherent—and therefore, more effective—resources for Response to Intervention (RTI), extra practice, and summer learning.

## Learn

*Eureka Math Learn* serves as a student's in-class companion where they show their thinking, share what they know, and watch their knowledge build every day. *Learn* assembles the daily classwork—Application Problems, Exit Tickets, Problem Sets, templates—in an easily stored and navigated volume.

## Practice

Each *Eureka Math* lesson begins with a series of energetic, joyous fluency activities, including those found in *Eureka Math Practice*. Students who are fluent in their math facts can master more material more deeply. With *Practice*, students build competence in newly acquired skills and reinforce previous learning in preparation for the next lesson.

Together, *Learn* and *Practice* provide all the print materials students will use for their core math instruction.

## Succeed

*Eureka Math Succeed* enables students to work individually toward mastery. These additional problem sets align lesson by lesson with classroom instruction, making them ideal for use as homework or extra practice. Each problem set is accompanied by a Homework Helper, a set of worked examples that illustrate how to solve similar problems.

Teachers and tutors can use *Succeed* books from prior grade levels as curriculum-consistent tools for filling gaps in foundational knowledge. Students will thrive and progress more quickly as familiar models facilitate connections to their current grade-level content.

# Students, families, and educators:

Thank you for being part of the *Eureka Math*™ community, where we celebrate the joy, wonder, and thrill of mathematics.

In the *Eureka Math* classroom, new learning is activated through rich experiences and dialogue. The *Learn* book puts in each student's hands the prompts and problem sequences they need to express and consolidate their learning in class.

## *What is in the* Learn *book?*

**Application Problems:** Problem solving in a real-world context is a daily part of *Eureka Math*. Students build confidence and perseverance as they apply their knowledge in new and varied situations. The curriculum encourages students to use the RDW process—Read the problem, Draw to make sense of the problem, and Write an equation and a solution. Teachers facilitate as students share their work and explain their solution strategies to one another.

**Problem Sets:** A carefully sequenced Problem Set provides an in-class opportunity for independent work, with multiple entry points for differentiation. Teachers can use the Preparation and Customization process to select "Must Do" problems for each student. Some students will complete more problems than others; what is important is that all students have a 10-minute period to immediately exercise what they've learned, with light support from their teacher.

Students bring the Problem Set with them to the culminating point of each lesson: the Student Debrief. Here, students reflect with their peers and their teacher, articulating and consolidating what they wondered, noticed, and learned that day.

**Exit Tickets:** Students show their teacher what they know through their work on the daily Exit Ticket. This check for understanding provides the teacher with valuable real-time evidence of the efficacy of that day's instruction, giving critical insight into where to focus next.

**Templates:** From time to time, the Application Problem, Problem Set, or other classroom activity requires that students have their own copy of a picture, reusable model, or data set. Each of these templates is provided with the first lesson that requires it.

## *Where can I learn more about* Eureka Math *resources?*

The Great Minds® team is committed to supporting students, families, and educators with an ever-growing library of resources, available at eureka-math.org. The website also offers inspiring stories of success in the *Eureka Math* community. Share your insights and accomplishments with fellow users by becoming a *Eureka Math* Champion.

Best wishes for a year filled with aha moments!

*Jill Diniz*

Jill Diniz
Director of Mathematics
Great Minds

# The Read–Draw–Write Process

The *Eureka Math* curriculum supports students as they problem-solve by using a simple, repeatable process introduced by the teacher. The Read–Draw–Write (RDW) process calls for students to

1. Read the problem.
2. Draw and label.
3. Write an equation.
4. Write a word sentence (statement).

Educators are encouraged to scaffold the process by interjecting questions such as

- What do you see?
- Can you draw something?
- What conclusions can you make from your drawing?

The more students participate in reasoning through problems with this systematic, open approach, the more they internalize the thought process and apply it instinctively for years to come.

# Contents

## Module 4: Angle Measure and Plane Figures

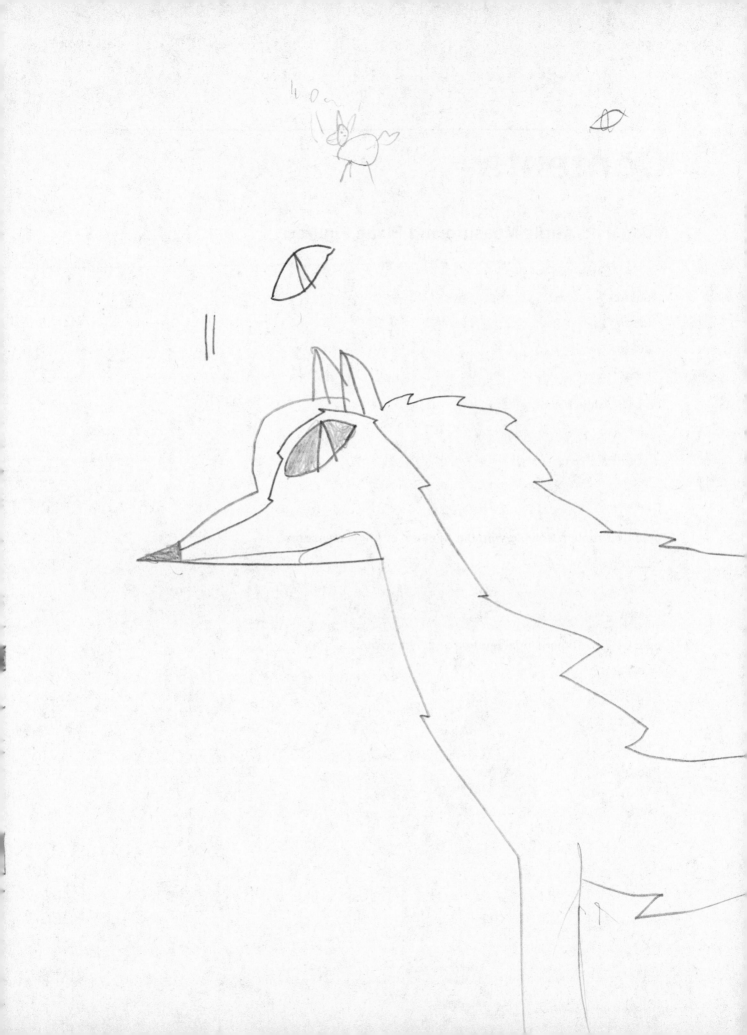

Name _____    Date _____

1. Use the following directions to draw a figure in the box to the right.

   a. Draw two points: $A$ and $B$.

   b. Use a straightedge to draw $\overleftrightarrow{AB}$.

   c. Draw a new point that is not on $\overleftrightarrow{AB}$. Label it $C$.

   d. Draw $\overline{AC}$.

   e. Draw a point not on $\overleftrightarrow{AB}$ or $\overline{AC}$. Call it $D$.

   f. Construct $\overleftrightarrow{CD}$.

   g. Use the points you've already labeled to name one angle. _____ C _____

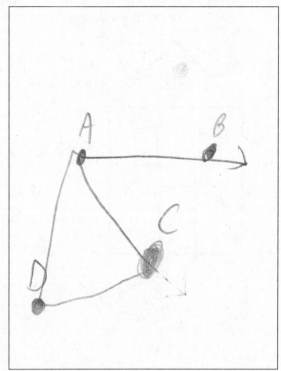

2. Use the following directions to draw a figure in the box to the right.

   a. Draw two points: $A$ and $B$.

   b. Use a straightedge to draw $\overline{AB}$.

   c. Draw a new point that is not on $\overline{AB}$. Label it $C$.

   d. Draw $\overrightarrow{BC}$.

   e. Draw a new point that is not on $\overline{AB}$ or $\overrightarrow{BC}$. Label it $D$.

   f. Construct $\overleftrightarrow{AD}$.

   g. Identify $\angle DAB$ by drawing an arc to indicate the position of the angle.

   h. Identify another angle by referencing points that you have already drawn. _____

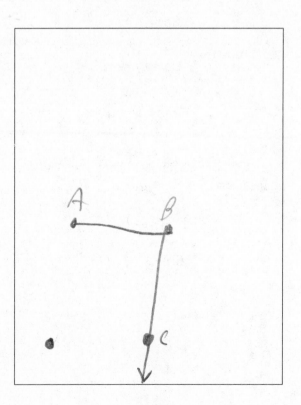

**EUREKA MATH**™    Lesson 1:    Identify and draw points, lines, line segments, rays, and angles. Recognize them in various contexts and familiar figures.     1

©2018 Great Minds®. eureka-math.org

3. a. Observe the familiar figures below. Label some points on each figure.

   b. Use those points to label and name representations of each of the following in the table below: ray, line, line segment, and angle. Extend segments to show lines and rays.

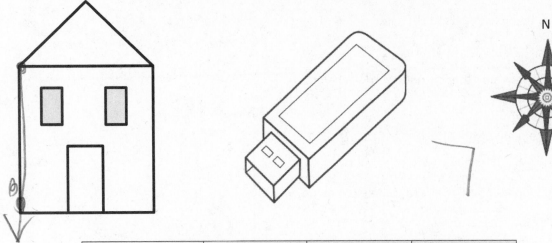

|  | House | Flash drive | Compass rose |
|---|---|---|---|
| Ray | $\overrightarrow{AB}$ |  |  |
| Line | $\overleftrightarrow{CB}$ |  |  |
| Line segment | $\overline{EF}$ |  |  |
| Angle | $\angle ABD$ |  |  |

Extension: Draw a familiar figure. Label it with points, and then identify rays, lines, line segments, and angles as applicable.

Lesson 1: Identify and draw points, lines, line segments, rays, and angles. Recognize them in various contexts and familiar figures.

©2018 Great Minds®. eureka-math.org

EUREKA MATH™

1. Figure 1 has three points. Connect points $A$, $B$, and $C$ with as many line segments as possible.

2. Figure 2 has four points. Connect points $D$, $E$, $F$, and $G$ with as many line segments as possible.

Figure 1

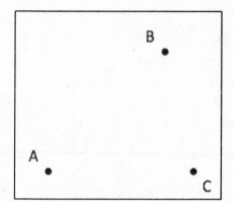

Figure 2

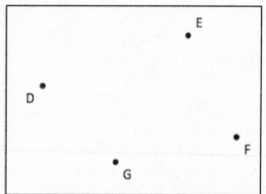

**Read          Draw          Write**

Lesson 2:   Use right angles to determine whether angles are equal to, greater than, or less than right angles. Draw right, obtuse, and acute angles.

5

©2018 Great Minds®. eureka-math.org

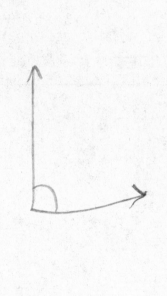

Name _____     Date _____

1. Use the right angle template that you made in class to determine if each of the following angles is greater than, less than, or equal to a right angle. Label each as *greater than*, *less than*, or *equal to*, and then connect each angle to the correct label of acute, right, or obtuse.
   The first one has been completed for you.

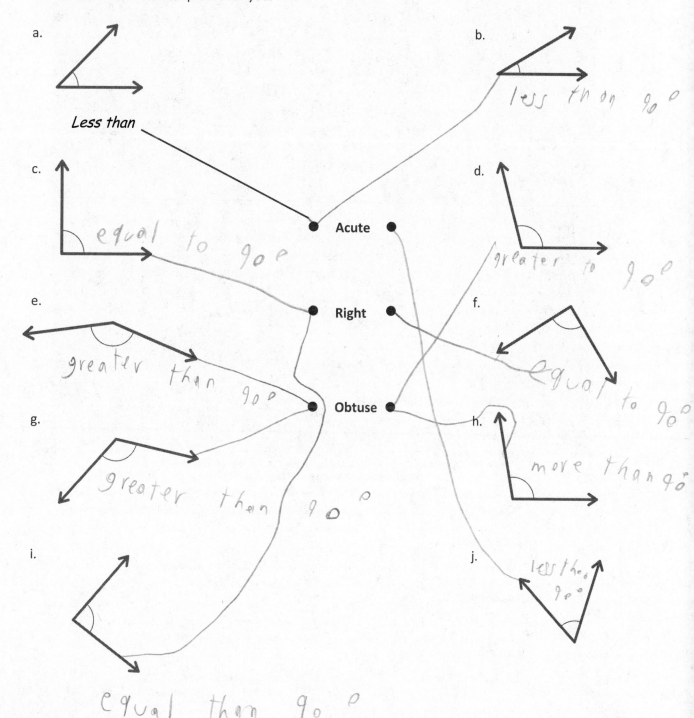

**EUREKA MATH**

**Lesson 2:**    Use right angles to determine whether angles are equal to, greater than, or less than right angles. Draw right, obtuse, and acute angles.

**7**

©2018 Great Minds®. eureka-math.org

2.  Use your right angle template to identify acute, obtuse, and right angles within Picasso's painting *Factory, Horta de Ebbo*. Trace at least two of each, label with points, and then name them in the table below the painting.

© 2013 Estate of Pablo Picasso / Artists Rights Society (ARS), New York
Photo: Erich Lessing / Art Resource, NY.

| Acute angle | ∠DCB | ∠DBC |
|---|---|---|
| Obtuse angle | ∠ABC | ∠JaC |
| Right angle | ∠DaB | ∠Xry |

**Lesson 2:**        Use right angles to determine whether angles are equal to, greater than, or less than right angles. Draw right, obtuse, and acute angles.

©2018 Great Minds®. eureka-math.org

EUREKA MATH™

3.  Construct each of the following using a straightedge and the right angle template that you created. Explain the characteristics of each by comparing the angle to a right angle. Use the words *greater than*, *less than,* or *equal to* in your explanations.

    a.  Acute angle

    b.  Right angle

    c.  Obtuse angle

**Lesson 2:**    Use right angles to determine whether angles are equal to, greater than, or less than right angles. Draw right, obtuse, and acute angles.

©2018 Great Minds®. eureka-math.org

9

Name _____     Date _____

1. Fill in the blanks to make true statements using one of the following words:  *acute, obtuse, right, straight.*

   a.  In class, we made a _____ angle when we folded paper twice.

   b.  An _____ angle is smaller than a right angle.

   c.  An _____ angle is larger than a right angle, but smaller than a straight angle.

2. Use a right angle template to identify the angles below.

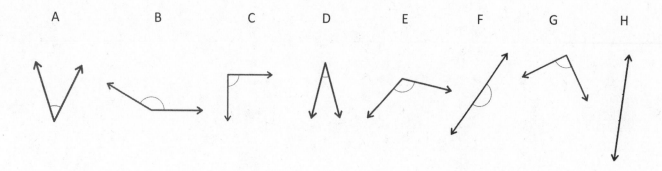

   A          B          C          D          E          F          G          H

   a.  Which angles are right angles?  _____

   b.  Which angles are obtuse angles?  _____

   c.  Which angles are acute angles?  _____

   d.  Which angles are straight angles?  _____

**EUREKA MATH™**

Lesson 2:     Use right angles to determine whether angles are equal to, greater than, or less than right angles.  Draw right, obtuse, and acute angles.

11

©2018 Great Minds®. eureka-math.org

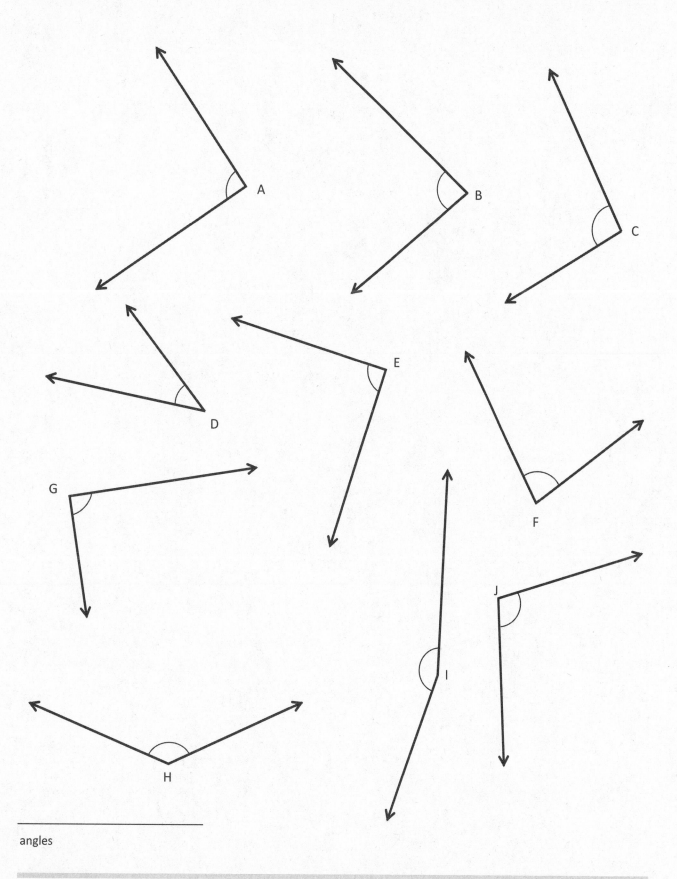

angles

Lesson 2:  Use right angles to determine whether angles are equal to, greater than, or less than right angles. Draw right, obtuse, and acute angles.

©2018 Great Minds®. eureka-math.org

13

a.  Estimate to draw point $X$ halfway up $\overline{AB}$.

b.  Estimate point $Y$ halfway up $\overline{CD}$.

c.  Draw horizontal line segment $XY$.  What word do the segments create?

d.  Erase segment $XY$.  Draw segment $CF$.  What word do the segments create?

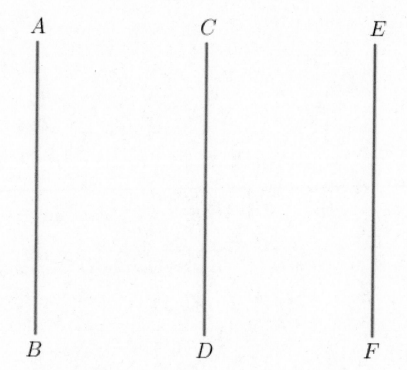

**Read**          **Draw**          **Write**

Name _____    Date _____

1. On each object, trace at least one pair of lines that appear to be perpendicular.

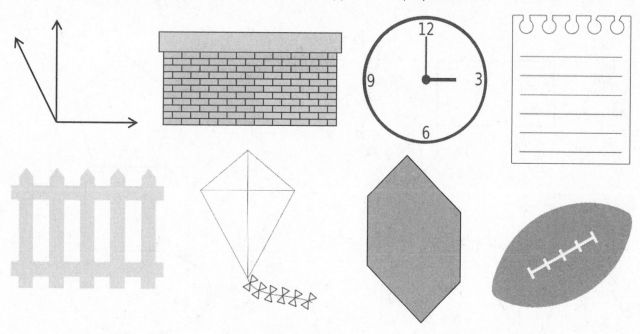

2. How do you know if two lines are perpendicular?




3. In the square and triangular grids below, use the given segments in each grid to draw a segment that is perpendicular using a straightedge.

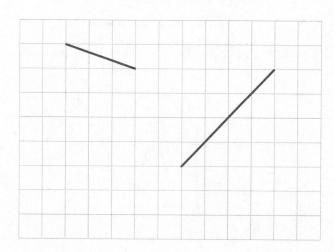 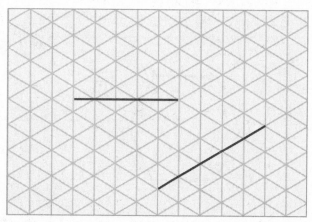

4. Use the right angle template that you created in class to determine which of the following figures have a right angle. Mark each right angle with a small square. For each right angle you find, name the corresponding pair of perpendicular sides. (Problem 4(a) has been started for you.)

a.

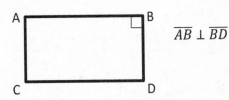

$$\overline{AB} \perp \overline{BD}$$

b.

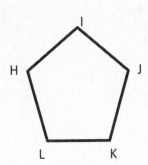

c.

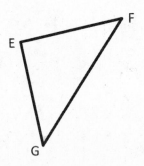

d.

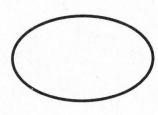

e.

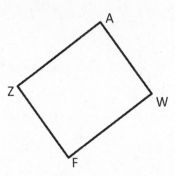

f.

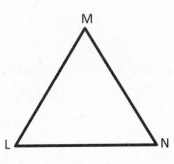

g.

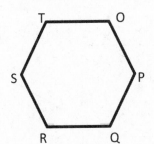

h.

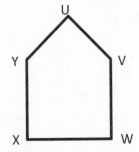

**EUREKA MATH**

5.  Mark each right angle on the following figure with a small square.  (Note:  A right angle does not have to be inside the figure.)  How many pairs of perpendicular sides does this figure have?

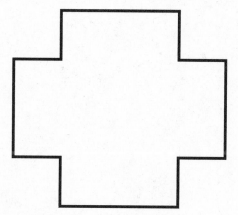

6.  True or false?  Shapes that have at least one right angle also have at least one pair of perpendicular sides. Explain your thinking.

Name _____     Date _____

Use a right angle template to measure the angles in the following figures.  Mark each right angle with a small square.  Then, name all pairs of perpendicular sides.

1.

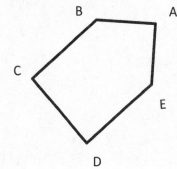

$\overline{BC}$ ⊥ _____

2.

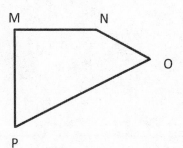

$\overline{MN}$ ⊥ _____

Observe the letters $R$, $E$, $A$, and $L$.

# R E A L

a. How many lines are perpendicular? Describe them.

b. How many acute angles are there? Describe them.

c. How many obtuse angles are there? Describe them.

**Read          Draw          Write**

Name _____ Date _____

1. On each object, trace at least one pair of lines that appear to be parallel.

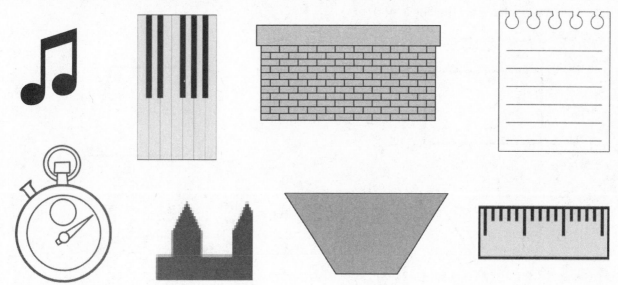

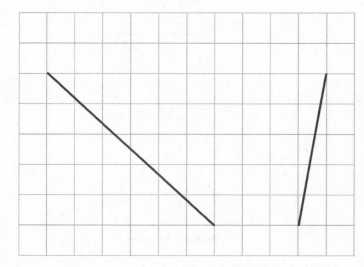

2. How do you know if two lines are parallel?

3. In the square and triangular grids below, use the given segments in each grid to draw a segment that is parallel using a straightedge.

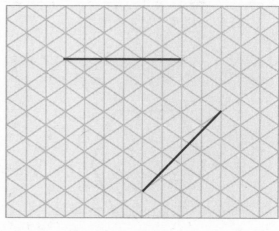

4. Determine which of the following figures have sides that are parallel by using a straightedge and the right angle template that you created. Circle the letter of the shapes that have at least one pair of parallel sides. Mark each pair of parallel sides with arrowheads, and then identify the parallel sides with a statement modeled after the one in 4(a).

a.

$$\overline{AB} \parallel \overline{CD}$$

b.

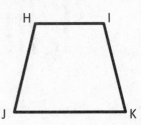

c.

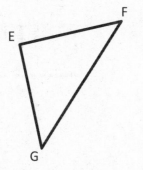

d.

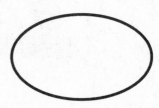

e.

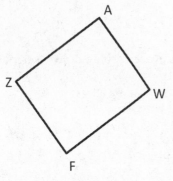

f.

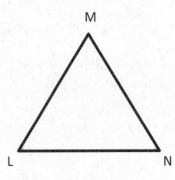

g.

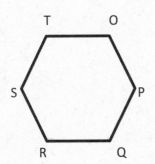

h.

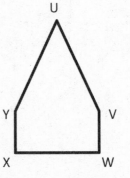

EUREKA
MATH™

5. True or false? A triangle cannot have sides that are parallel. Explain your thinking.

6. Explain why $\overline{AB}$ and $\overline{CD}$ are parallel, but $\overline{EF}$ and $\overline{GH}$ are not.

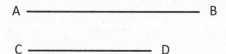
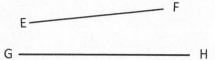

7. Draw a line using your straightedge. Now, use your right angle template and straightedge to construct a line parallel to the first line you drew.

EUREKA
MATH™

Name _____     Date _____

Look at the following pairs of lines.  Identify if they are parallel, perpendicular, or intersecting.

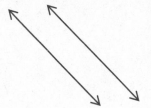

1. _____

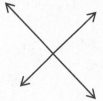

2. _____

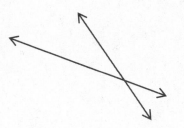

3. _____

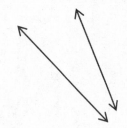

4. _____

Place right angle templates on top of the circle to determine how many right angles can fit around the center point of the circle.  (Overlaps are not allowed.)  How many right angles can fit?

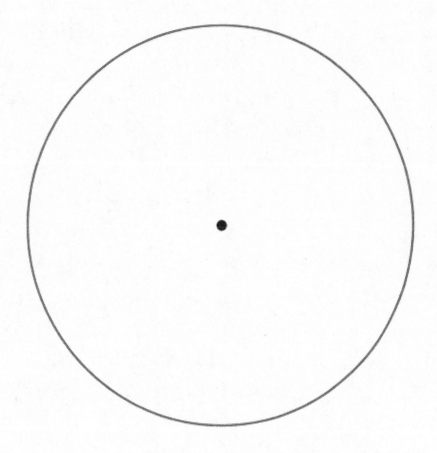

Read          Draw          Write

**Lesson 5:**     Use a circular protractor to understand a 1-degree angle as $\frac{1}{360}$ of a
turn.  Explore benchmark angles using the protractor.

31

©2018 Great Minds®. eureka-math.org

Name _____  Date _____

1.  Make a list of the measures of the benchmark angles you drew, starting with Set A.
    Round each angle measure to the nearest 5°.  Both sets have been started for you.

    a.  Set A:  45°, 90°,

    b.  Set B:  30°, 60°,

2.  Circle any angle measures that appear on both lists.  What do you notice about them?

3.  List the angle measures from Problem 1 that are acute.  Trace each angle with your finger as you say its measurement.

4.  List the angle measures from Problem 1 that are obtuse.  Trace each angle with your finger as you say its measurement.

Lesson 5:     Use a circular protractor to understand a 1-degree angle as $\frac{1}{360}$ of a turn.  Explore benchmark angles using the protractor.

©2018 Great Minds®. eureka-math.org

33

5.  We found out today that $1°$ is $\frac{1}{360}$ of a whole turn.  It is 1 out of 360°.  That means a 2° angle is $\frac{2}{360}$ of a whole turn.  What fraction of a whole turn is each of the benchmark angles you listed in Problem 1?

6.  How many 45° angles does it take to make a full turn?

7.  How many 30° angles does it take to make a full turn?

8.  If you didn't have a protractor, how could you reconstruct a quarter of it from 0° to 90°?

**Lesson 5:**     Use a circular protractor to understand a 1-degree angle as $\frac{1}{360}$ of a turn.  Explore benchmark angles using the protractor.

©2018 Great Minds®. eureka-math.org

Name _____ Date _____

1. How many right angles make a full turn?

2. What is the measurement of a right angle?

3. What fraction of a full turn is 1°?

4. Name at least four benchmark angle measurements.

Lesson 5:   ·  Use a circular protractor to understand a 1-degree angle as $\frac{1}{360}$ of a turn. Explore benchmark angles using the protractor.

©2018 Great Minds®. eureka-math.org

35

Cut out the circles on the template on the next page. Fold Circle A and Circle B as you would to make a right angle template. Trace the folded perpendicular lines. How many right angles do you see at the center of each circle? Did the size of the circle matter?

_____

_____

_____

_____

**Read**          **Draw**          **Write**

EUREKA
MATH

Lesson 6:     Use varied protractors to distinguish angle measure from length
              measurement.

37

©2018 Great Minds®. eureka-math.org

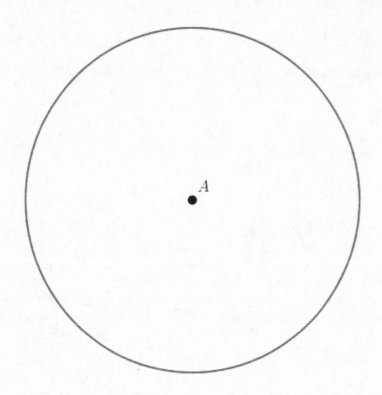

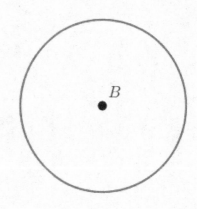

**Read**      **Draw**      **Write**

Lesson 6:    Use varied protractors to distinguish angle measure from length measurement.

©2018 Great Minds®. eureka-math.org

39

Name _____     Date _____

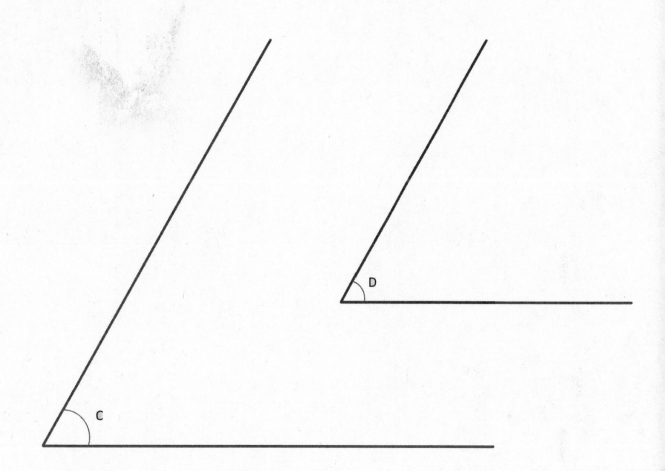

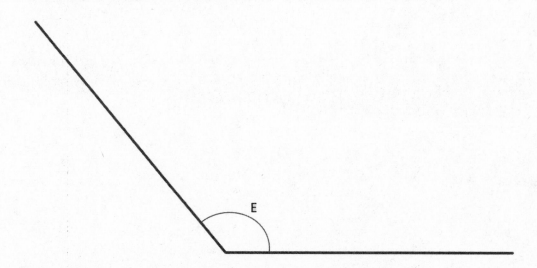

**EUREKA MATH**™

**Lesson 6:**    Use varied protractors to distinguish angle measure from length
measurement.

**41**

©2018 Great Minds®. eureka-math.org

Name _____    Date _____

1.  Use a protractor to measure the angles, and then record the measurements in degrees.

    a.                                              b.

    c.                                              d.

**Lesson 6:**    Use varied protractors to distinguish angle measure from length
measurement.

43

©2018 Great Minds®. eureka-math.org

e.

f.

g.

h.

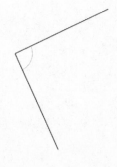

i.

j.

**Lesson 6:**     Use varied protractors to distinguish angle measure from length measurement.

©2018 Great Minds®. eureka-math.org

EUREKA
MATH™

2. a.  Use three different-size protractors to measure the angle.  Extend the lines as needed using a straightedge.

Protractor #1: _____ °

Protractor #2: _____ °

Protractor #3: _____ °

b.  What do you notice about the measurement of the above angle using each of the protractors?

3. Use a protractor to measure each angle.  Extend the length of the segments as needed.  When you extend the segments, does the angle measure stay the same?  Explain how you know.

a.

b.

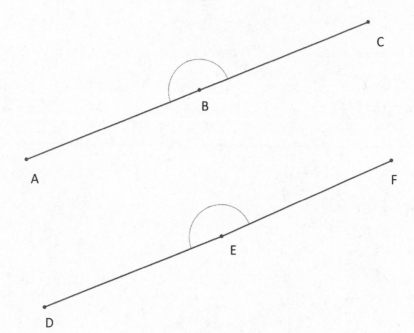

Lesson 6:     Use varied protractors to distinguish angle measure from length measurement.

45

©2018 Great Minds®. eureka-math.org

Name _____     Date _____

Use any protractor to measure the angles, and then record the measurements in degrees.

1.

2.

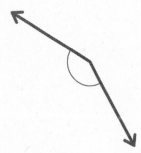

3.

4.

EUREKA MATH™

Lesson 6:     Use varied protractors to distinguish angle measure from length measurement.

©2018 Great Minds®. eureka-math.org

47

Predict the measure of ∠$XYZ$ using your right angle template.  Then, find the actual measure of

∠$XYZ$ using a circular protractor and 180° protractor.

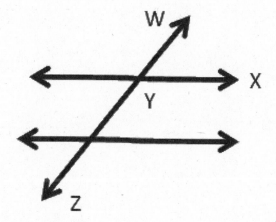

Lesson 7:      Measure and draw angles.  Sketch given angle measures, and verify
         with a protractor.

49

©2018 Great Minds®. eureka-math.org

**Read**        **Draw**        **Write**

Name _____    Date _____

Figure 1

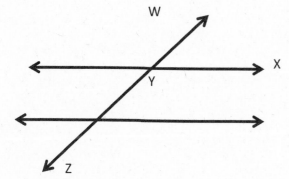

Figure 2

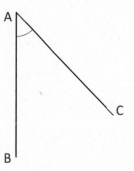

Figure 3

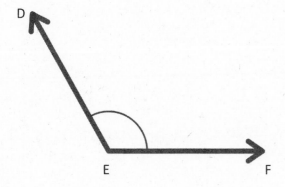

Figure 4

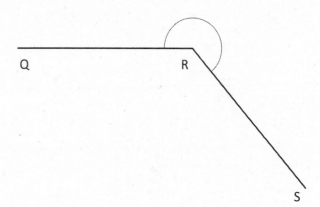

EUREKA MATH    **Lesson 7:**    Measure and draw angles.  Sketch given angle measures, and verify    51
with a protractor.

©2018 Great Minds®. eureka-math.org

Name _____    Date _____

Construct angles that measure the given number of degrees. For Problems 1–4, use the ray shown as one of the rays of the angle with its endpoint as the vertex of the angle. Draw an arc to indicate the angle that was measured.

1.  30°                                             2.  65°

3.  115°                                            4.  135°

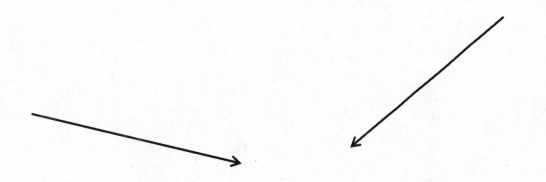

**EUREKA MATH**

Lesson 7:      Measure and draw angles. Sketch given angle measures, and verify
              with a protractor.

53

©2018 Great Minds®. eureka-math.org

5. 5°                                        6. 175°

7. 27°                                       8. 117°

9. 48°                                       10. 132°

Lesson 7:    Measure and draw angles. Sketch given angle measures, and verify
             with a protractor.

EUREKA
MATH™

Name _____ Date _____

Construct angles that measure the given number of degrees. Draw an arc to indicate the angle that was measured.

1. 75°

2. 105°

3. 81°

4. 99°

Lesson 7:   Measure and draw angles. Sketch given angle measures, and verify
             with a protractor.

©2018 Great Minds®. eureka-math.org

55

Draw a series of clocks that show 12:00, 3:00, 6:00, and 9:00.  Use an arc to identify an angle and estimate the angle created by both hands on the clock.

_____

_____

_____

_____

**Read**          **Draw**          **Write**

**Lesson 8:**    Identify and measure angles as turns and recognize them in various contexts.

57

©2018 Great Minds®. eureka-math.org

Name _____   Date _____

1. Joe, Steve, and Bob stood in the middle of the yard and faced the house. Joe turned 90° to the right. Steve turned 180° to the right. Bob turned 270° to the right. Name the object that each boy is now facing.

   Joe _____

   Steve _____

   Bob _____

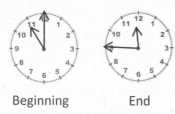

2. Monique looked at the clock at the beginning of class and at the end of class. How many degrees did the minute hand turn from the beginning of class until the end?

   Beginning    End

3. The skater jumped into the air and did a 360. What does that mean?

4. Mr. Martin drove away from his house without his wallet. He did a 180. Where is he heading now?

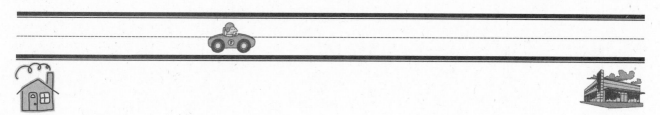

House                                                    Store

Lesson 8:   Identify and measure angles as turns and recognize them in various contexts.

©2018 Great Minds®. eureka-math.org

59

5. John turned the knob of the shower 270° to the right. Draw a picture showing the position of the knob after he turned it.

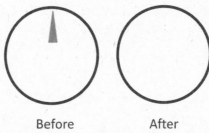

Before          After

6. Barb used her scissors to cut out a coupon from the newspaper. How many quarter-turns does she need to turn the paper in order to stay on the lines?

7. How many quarter-turns does the picture need to be rotated in order for it to be upright?

8. Meredith faced north. She turned 90° to the right, and then 180° more. In which direction is she now facing?

Lesson 8:   Identify and measure angles as turns and recognize them in various contexts.

©2018 Great Minds®. eureka-math.org

EUREKA MATH™

Name _____     Date _____

1.  Marty was doing a handstand.  Describe how many degrees his body will turn to be upright again.

2.  Jeffrey started riding his bike at the .  He travelled north for 3 blocks, then turned 90° to the right and rode for 2 blocks.  In which direction was he headed?  Sketch his route on the grid below.  Each square unit represents 1 block.

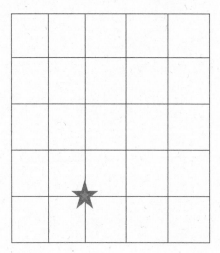

EUREKA
MATH™

Lesson 8:  Identify and measure angles as turns and recognize them in various contexts.

©2018 Great Minds®. eureka-math.org

61

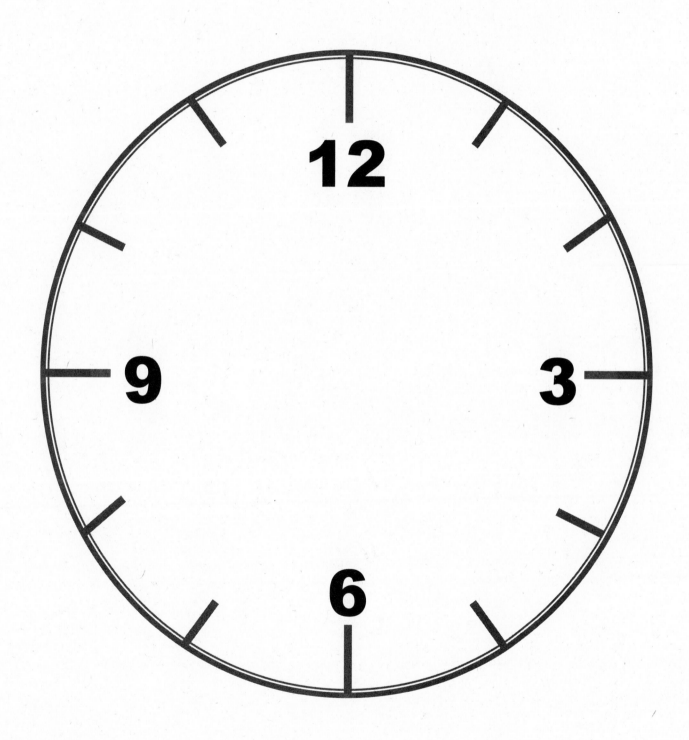

clock

Lesson 8:     Identify and measure angles as turns and recognize them in various contexts.

©2018 Great Minds®. eureka-math.org

63

List times on the clock in which the angle between the hour and minute hands is 90°.  Verify your work using a protractor.

Stay alert for this misconception: Why don't the hands at 3:30 form a 90° angle as expected?

_____

_____

_____

_____

**Read**         **Draw**         **Write**

Name _____ Date _____

1. Complete the table.

| Pattern block | Total number that fit around 1 vertex | One interior angle measures... | Sum of the angles around a vertex |
|---|---|---|---|
| a. | | $360° ÷$ ____ $=$ ____ | ____ $+$ ____ $+$ ____ $+$ ____ $= 360°$ |
| b. | | | |
| c. | | | ____ $+$ ____ $+$ ____ $= 360°$ |
| d. (Acute angle) | | | |
| e. (Obtuse angle) | | | |
| f. (Acute angle) | | | |

2. Find the measurements of the angles indicated by the arcs.

| Pattern blocks | Angle measure | Addition sentence |
|---|---|---|
| a. | | |
| b. | | |
| c. | | |

3. Use two or more pattern blocks to figure out the measurements of the angles indicated by the arcs.

| Pattern blocks | Angle measure | Addition sentence |
|---|---|---|
| a. | | |
| b. | | |
| c. | | |

Lesson 9:    Decompose angles using pattern blocks.

Name _____ Date _____

1. Describe and sketch two combinations of the blue rhombus pattern block that create a straight angle.

2. Describe and sketch two combinations of the green triangle and yellow hexagon pattern block that create a straight angle.

Using pattern blocks of the same shape or different shapes, construct a straight angle.  Which shapes did you use?  Which pattern block can you add to your existing shape to create a 270° angle? How can you tell?

_____

_____

_____

_____

**Read          Draw          Write**

EUREKA
MATH™

**Lesson 10:**     Use the addition of adjacent angle measures to solve problems using a
symbol for the unknown angle measure.

71

©2018 Great Minds®. eureka-math.org

Name _____ Date _____

Write an equation, and solve for the measure of $\angle x$. Verify the measurement using a protractor.

1. $\angle CBA$ is a right angle.

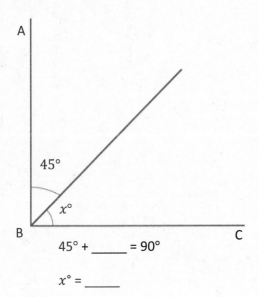

45° + _____ = 90°

$x° =$ _____

2. $\angle GFE$ is a right angle.

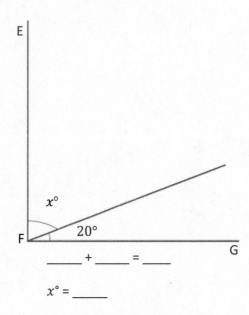

_____ + _____ = _____

$x° =$ _____

3. $\angle IJK$ is a straight angle.

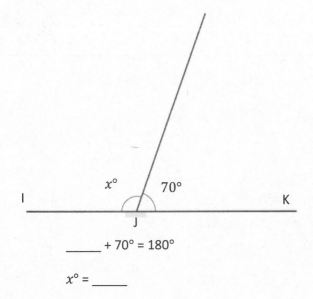

_____ + 70° = 180°

$x° =$ _____

4. $\angle MNO$ is a straight angle.

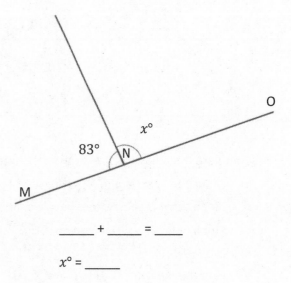

_____ + _____ = _____

$x° =$ _____

EUREKA MATH

**Lesson 10:** Use the addition of adjacent angle measures to solve problems using a symbol for the unknown angle measure.

73

©2018 Great Minds®. eureka-math.org

Solve for the unknown angle measurements. Write an equation to solve.

5. Solve for the measurement of ∠TRU.
   ∠QRS is a straight angle.

6. Solve for the measurement of ∠ZYV.
   ∠XYZ is a straight angle.

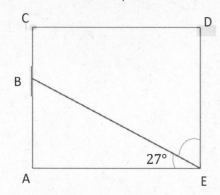

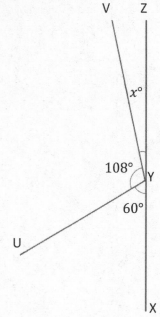

7. In the following figure, ACDE is a rectangle. Without using a protractor, determine the measurement of ∠DEB. Write an equation that could be used to solve the problem.

8. Complete the following directions in the space to the right.

   a. Draw 2 points: M and N. Using a straightedge, draw $\overleftrightarrow{MN}$.

   b. Plot a point O somewhere between points M and N.

   c. Plot a point P, which is not on $\overleftrightarrow{MN}$.

   d. Draw $\overline{OP}$.

   e. Find the measure of ∠MOP and ∠NOP.

   f. Write an equation to show that the angles add to the measure of a straight angle.

Lesson 10:    Use the addition of adjacent angle measures to solve problems using a symbol for the unknown angle measure.

©2018 Great Minds®. eureka-math.org

Name _____    Date _____

Write an equation, and solve for $x$. $\angle TUV$ is a straight angle.

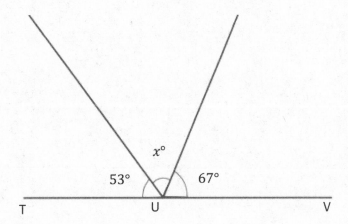

Equation: _____

$x° = $ _____

EUREKA
MATH™

Lesson 10:    Use the addition of adjacent angle measures to solve problems using a
symbol for the unknown angle measure.

©2018 Great Minds®. eureka-math.org

75

Use patterns blocks of various types to create a design in which you can see a decomposition of 360°. Which shapes did you use? Write an equation to show how you composed 360°.

_____

_____

_____

_____

**Read**          **Draw**          **Write**

Lesson 11:     Use the addition of adjacent angle measures to solve problems using a
              symbol for the unknown angle measure.

77

©2018 Great Minds®. eureka-math.org

Name _____     Date _____

Write an equation, and solve for the unknown angle measurements numerically.

1.

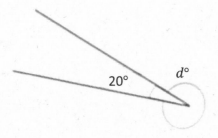

_____° + 20° = 360°

d° = _____°

2.

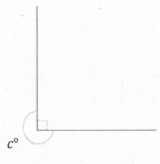

_____° + _____° = 360°

c° = _____°

3.

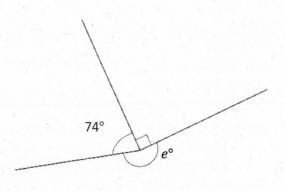

_____° + _____° + _____° = _____°

e° = _____°

4.

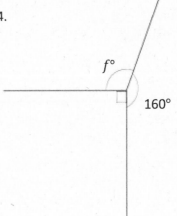

_____° + _____° + _____° = _____°

f° = _____°

EUREKA MATH

Lesson 11:   Use the addition of adjacent angle measures to solve problems using a symbol for the unknown angle measure.

©2018 Great Minds®. eureka-math.org

79

Write an equation, and solve for the unknown angles numerically.

5.  $O$ is the intersection of $\overline{AB}$ and $\overline{CD}$.
    $\angle DOA$ is 160°, and $\angle AOC$ is 20°.

    $x° = \underline{\hspace{1.5cm}}$  $y° = \underline{\hspace{1.5cm}}$

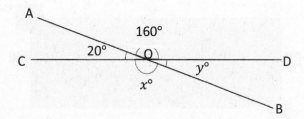

6.  $O$ is the intersection of $\overline{RS}$ and $\overline{TV}$.
    $\angle TOS$ is 125°.

    $g° = \underline{\hspace{1.5cm}}$  $h° = \underline{\hspace{1.5cm}}$  $i° = \underline{\hspace{1.5cm}}$

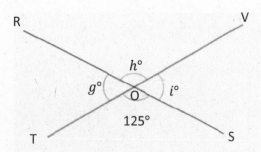

7.  $O$ is the intersection of $\overline{WX}$, $\overline{YZ}$, and $\overline{UO}$.
    $\angle XOZ$ is 36°.

    $k° = \underline{\hspace{1.5cm}}$  $m° = \underline{\hspace{1.5cm}}$  $n° = \underline{\hspace{1.5cm}}$

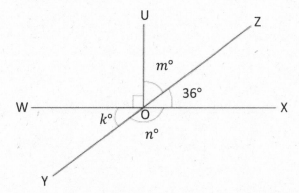

Lesson 11:    Use the addition of adjacent angle measures to solve problems using a
              symbol for the unknown angle measure.

EUREKA
MATH

Name _____     Date _____

Write equations using variables to represent the unknown angle measurements. Find the unknown angle measurements numerically.

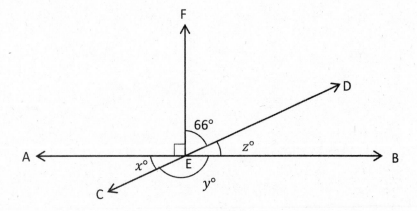

1.  $x° =$

2.  $y° =$

3.  $z° =$

**EUREKA MATH**™

**Lesson 11:**   Use the addition of adjacent angle measures to solve problems using a symbol for the unknown angle measure.

81

©2018 Great Minds®. eureka-math.org

Cut along the dotted line in the template on the next page, and unfold the figure. Notice how each side of the folded line matches. Fold another way, and see if the sides match. Observe the attributes of the figure and write a summary of your observations.

_____

_____

_____

_____

_____

_____

_____

_____

_____

_____

**Read        Draw        Write**

Lesson 12:    Recognize lines of symmetry for given two-dimensional figures.
              Identify line-symmetric figures, and draw lines of symmetry.

83

©2018 Great Minds®. eureka-math.org

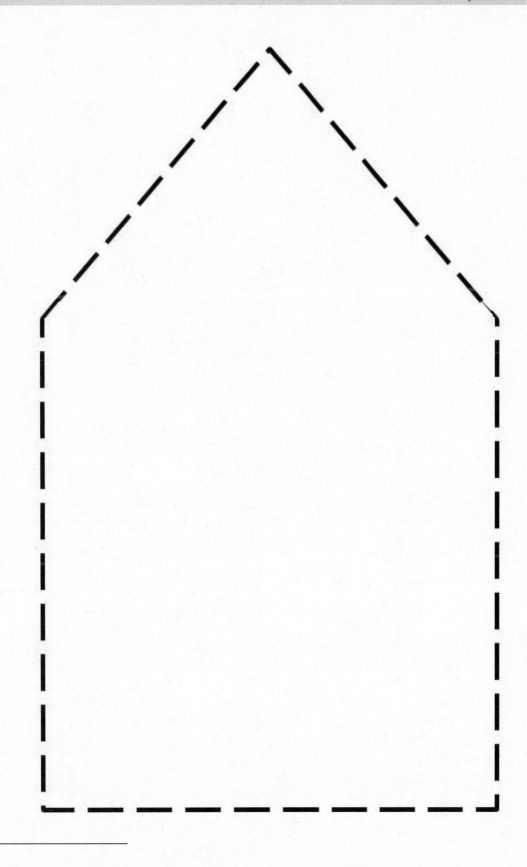

pentagon

Lesson 12:      Recognize lines of symmetry for given two-dimensional figures.
             Identify line-symmetric figures, and draw lines of symmetry.

85

Name _____  Date _____

1. Circle the figures that have a correct line of symmetry drawn.

a.   b. 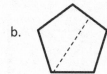  c.   d.

2. Find and draw all lines of symmetry for the following figures. Write the number of lines of symmetry that you found in the blank underneath the shape.

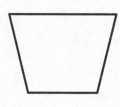

a. _____

b. _____

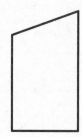

c. _____

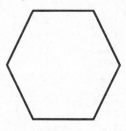

d. _____

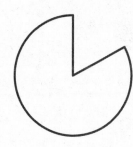

e. _____

f. _____

g. _____

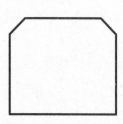

h. _____

i. _____

EUREKA MATH

Lesson 12:   Recognize lines of symmetry for given two-dimensional figures.
Identify line-symmetric figures, and draw lines of symmetry.

87

3. Half of each figure below has been drawn. Use the line of symmetry, represented by the dashed line, to complete each figure.

a.

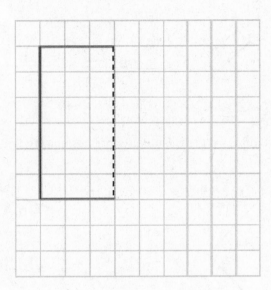

b.

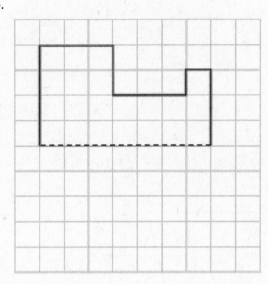

c.

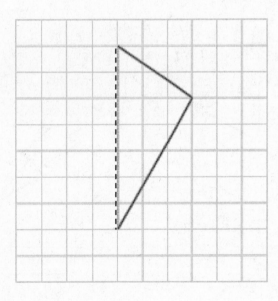

d.

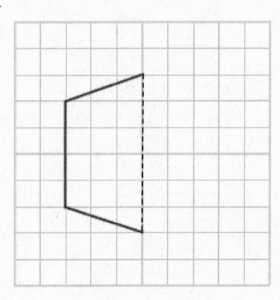

4. The figure below is a circle. How many lines of symmetry does the figure have? Explain.

Lesson 12: Recognize lines of symmetry for given two-dimensional figures. Identify line-symmetric figures, and draw lines of symmetry.

Name _____   Date _____

1.  Is the line drawn a line of symmetry?  Circle your choice.

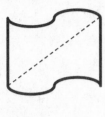          

Yes      No              Yes      No              Yes      No

2.  Draw as many lines of symmetry as you can find in the figure below.

EUREKA
MATH™

Lesson 12:   Recognize lines of symmetry for given two-dimensional figures.
Identify line-symmetric figures, and draw lines of symmetry.

89

©2018 Great Minds®. eureka-math.org

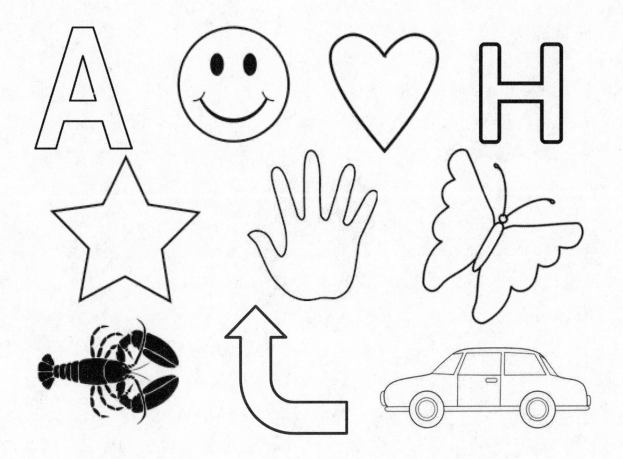

Figure 1

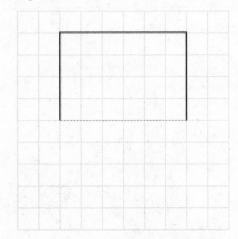

Figure 2

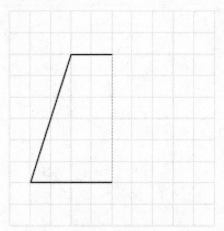

_____
lines of symmetry

        Lesson 12:    Recognize lines of symmetry for given two-dimensional figures.
                              Identify line-symmetric figures, and draw lines of symmetry.

©2018 Great Minds®. eureka-math.org

91

Cut along the dotted line in the template on the next page. Fold Triangles A, B, and C to show their lines of symmetry. Use a straightedge to trace each fold. Observe the relationships of symmetric shapes to angles and side lengths. Write a summary of your observations.

_____

_____

_____

_____

_____

_____

_____

_____

_____

**Read**          **Draw**          **Write**

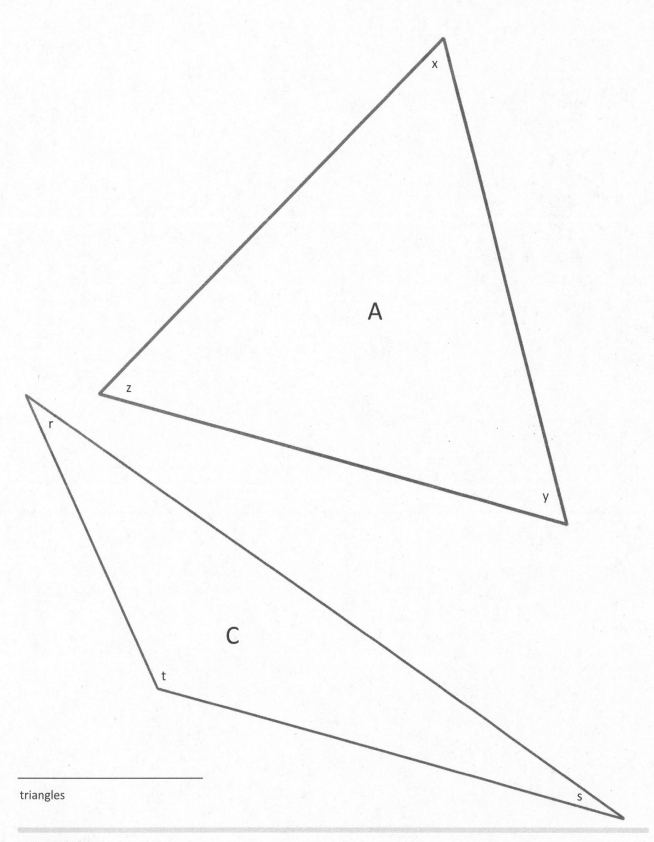

triangles

Lesson 13: Analyze and classify triangles based on side length, angle measure, or both.

©2018 Great Minds®. eureka-math.org

95

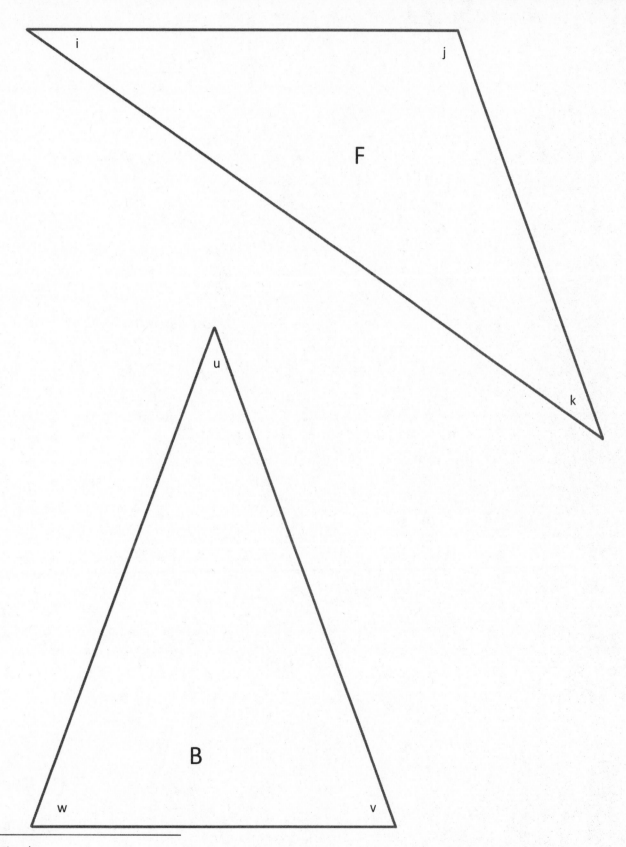

F

i

j

k

u

B

w

v

triangles

Lesson 13:    Analyze and classify triangles based on side length, angle measure, or both.

97

©2018 Great Minds®. eureka-math.org

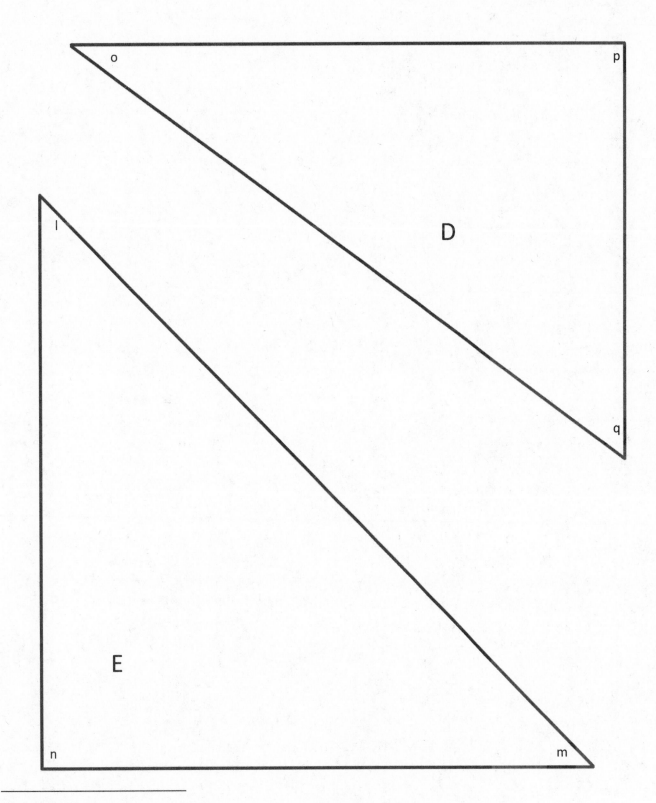

triangles

**Lesson 13:**  Analyze and classify triangles based on side length, angle measure, or both.

99

©2018 Great Minds®. eureka-math.org

Name _____  Date _____

| Sketch of Triangle | Attributes (Include side lengths and angle measures.) | Classification | |
|---|---|---|---|
| A | | | |
| B | | | |
| C | | | |
| D | | | |
| E | | | |
| F | | | |

**Lesson 13:** Analyze and classify triangles based on side length, angle measure, or both.

©2018 Great Minds®. eureka-math.org

**101**

Name _____     Date _____

1.  Classify each triangle by its side lengths and angle measurements.  Circle the correct names.

|  | Classify Using Side Lengths | Classify Using Angle Measurements |
|---|---|---|
| a. 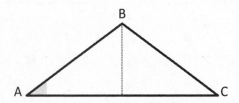 | Equilateral   Isosceles   Scalene | Acute   Right   Obtuse |
| b. | Equilateral   Isosceles   Scalene | Acute   Right   Obtuse |
| c. | Equilateral   Isosceles   Scalene | Acute   Right   Obtuse |
| d. | Equilateral   Isosceles   Scalene | Acute   Right   Obtuse |

2.  △ $ABC$ has one line of symmetry as shown.  What does this tell you about the measures of ∠$A$ and ∠$C$?

B

A           C

3.  △ $DEF$ has three lines of symmetry as shown.

a.  How can the lines of symmetry help you to figure out which angles are equal?

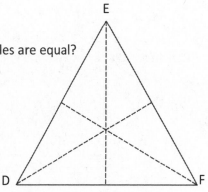

b.  △ $DEF$ has a perimeter of 30 cm.  Label the side lengths.

©2018 Great Minds®. eureka-math.org

4.  Use a ruler to connect points to form two other triangles. Use each point only once. None of the triangles may overlap. One or two points will be unused. Name and classify the three triangles below. The first one has been done for you.

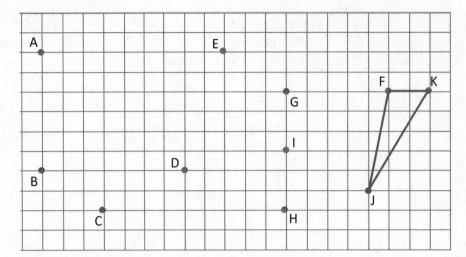

| Name the Triangles Using Vertices | Classify by Side Length | Classify by Angle Measurement |
|---|---|---|
| $\triangle FJK$ | Scalene | Obtuse |
| | | |
| | | |

5.  a.  List three points from the grid above that, when connected by segments, do not result in a triangle.

    b.  Why didn't the three points you listed result in a triangle when connected by segments?

6.  Can a triangle have two right angles? Explain.

**Lesson 13:**      Analyze and classify triangles based on side length, angle measure, or both.

EUREKA
MATH™

Name _____  Date _____

Use appropriate tools to solve the following problems.

1.  The triangles below have been classified by shared attributes (side length *or* angle type). Use the words *acute, right, obtuse, scalene, isosceles,* or *equilateral* to label the headings to identify the way the triangles have been sorted.

2.  Draw lines to identify each triangle according to angle type *and* side length.

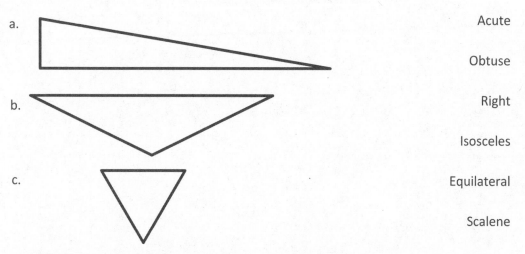

a.                                                    Acute

                                                     Obtuse

b.                                                    Right

                                                     Isosceles

c.                                                    Equilateral

                                                     Scalene

3.  Identify and draw any lines of symmetry in the triangles in Problem 2.

EUREKA
MATH™

**Lesson 13:**    Analyze and classify triangles based on side length, angle measure, or both.

**105**

©2018 Great Minds®. eureka-math.org

Draw three points on your grid paper so that, when connected, they form a triangle. Use your straightedge to connect the three points to form a triangle. Determine how the triangle you constructed can be classified: right, acute, obtuse, equilateral, isosceles, or scalene.

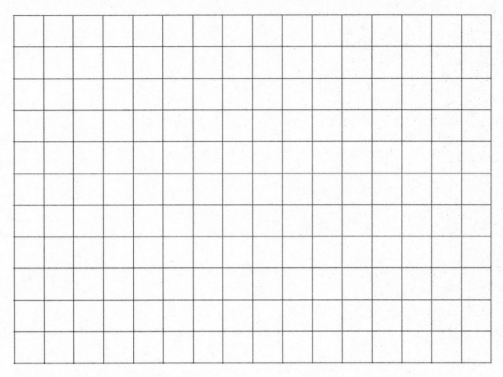

a. How can you classify your triangle?

**Read**        **Draw**        **Write**

**Lesson 14:**    Define and construct triangles from given criteria. Explore symmetry in triangles.

©2018 Great Minds®. eureka-math.org

**107**

b.  What attributes did you look at to classify the triangle?

c.  What tools did you use to help draw and classify your triangle?

**Read**          **Draw**          **Write**

**Lesson 14:**    Define and construct triangles from given criteria.  Explore symmetry
in triangles.

EUREKA
MATH™

Name _____  Date _____

1. Draw triangles that fit the following classifications. Use a ruler and protractor. Label the side lengths and angles.

   a.  Right and isosceles

   b.  Obtuse and scalene

   c.  Acute and scalene

   d.  Acute and isosceles

2. Draw all possible lines of symmetry in the triangles above. Explain why some of the triangles do not have lines of symmetry.

Lesson 14:    Define and construct triangles from given criteria. Explore symmetry in triangles.

109

©2018 Great Minds®. eureka-math.org

Are the following statements true or false?  Explain using pictures or words.

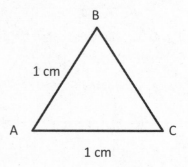

3.  If △ *ABC* is an equilateral triangle, $\overline{BC}$ must be 2 cm.  True or False?

4.  A triangle cannot have one obtuse angle and one right angle.  True or False?

5.  △ *EFG* can be described as a right triangle and an isosceles triangle.  True or False?

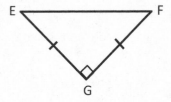

6.  An equilateral triangle is isosceles.  True or False?

Extension:  In △ *HIJ*,  a = b.  True or False?

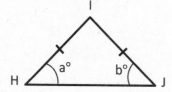

**Lesson 14:**     Define and construct triangles from given criteria.  Explore symmetry
                        in triangles.

                        ©2018 Great Minds®. eureka-math.org

Name _____ Date _____

1.  Draw an obtuse isosceles triangle, and then draw any lines of symmetry if they exist.

2.  Draw a right scalene triangle, and then draw any lines of symmetry if they exist.

3.  Every triangle has at least _____ acute angles.

Lesson 14:    Define and construct triangles from given criteria. Explore symmetry
              in triangles.

©2018 Great Minds®. eureka-math.org

111

a. On grid paper, draw two perpendicular line segments, each measuring 4 units, which extend from a point $V$. Identify the segments as $\overline{SV}$ and $\overline{UV}$. Draw $\overline{SU}$. What shape did you construct? Classify it.

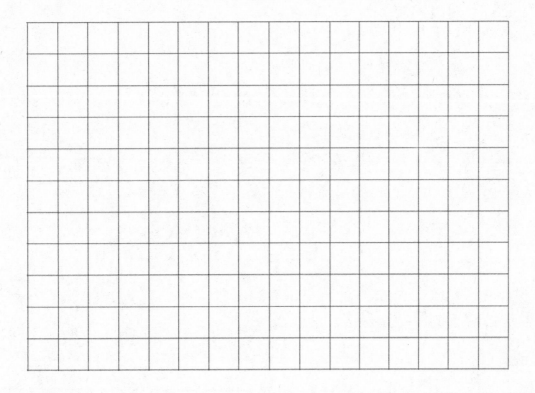

b. Imagine $\overline{SU}$ is a line of symmetry. Construct the other half of the figure. What figure did you construct? How can you tell?

**Read**          **Draw**          **Write**

Lesson 15:    Classify quadrilaterals based on parallel and perpendicular lines and the presence or absence of angles of a specified size.

©2018 Great Minds®. eureka-math.org

113

Name _____  Date _____

Construct the figures with the given attributes. Name the shape you created. Be as specific as possible. Use extra blank paper as needed.

1. Construct quadrilaterals with at least one set of parallel sides.

2. Construct a quadrilateral with two sets of parallel sides.

3. Construct a parallelogram with four right angles.

4. Construct a rectangle with all sides the same length.

Lesson 15:   Classify quadrilaterals based on parallel and perpendicular lines and
the presence or absence of angles of a specified size.

115

©2018 Great Minds®. eureka-math.org

5. Use the word bank to name each shape, being as specific as possible.

| Parallelogram | Trapezoid | Rectangle | Square |
|---|---|---|---|

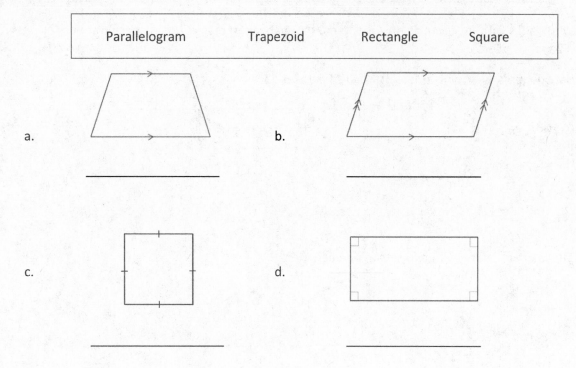

a. _____

b. _____

c. _____

d. _____

6. Explain the attribute that makes a square a special rectangle.

7. Explain the attribute that makes a rectangle a special parallelogram.

8. Explain the attribute that makes a parallelogram a special trapezoid.

Lesson 15: Classify quadrilaterals based on parallel and perpendicular lines and the presence or absence of angles of a specified size.

©2018 Great Minds®. eureka-math.org

EUREKA MATH

Name _____  Date _____

1.  In the space below, draw a parallelogram.

2.  Explain why a rectangle is a special parallelogram.

**Lesson 15:**    Classify quadrilaterals based on parallel and perpendicular lines and
the presence or absence of angles of a specified size.

**117**

©2018 Great Minds®. eureka-math.org

Within the stars, find at least two different examples for each of the following.  Explain which attributes you used to identify each.

- Equilateral triangles
- Trapezoids
- Parallelograms
- Rhombuses

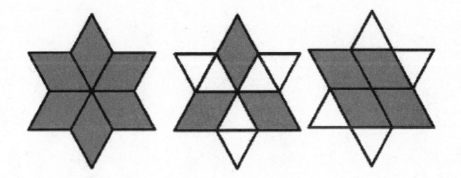

_____

_____

_____

_____

**Read**        **Draw**        **Write**

**Lesson 16:**    Reason about attributes to construct quadrilaterals on square or triangular grid paper.

**119**

©2018 Great Minds®. eureka-math.org

Name _____ Date _____

1. On the grid paper, draw at least one quadrilateral to fit the description. Use the given segment as one segment of the quadrilateral. Name the figure you drew using one of the terms below.

| Parallelogram | Trapezoid | Rectangle |
|---|---|---|
| Square | | Rhombus |

a. A quadrilateral that has at least one pair of parallel sides.

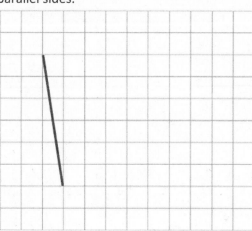

b. A quadrilateral that has four right angles.

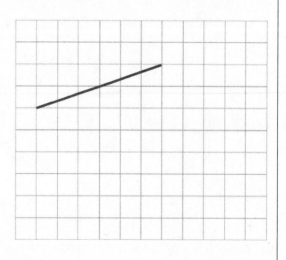

c. A quadrilateral that has two pairs of parallel side

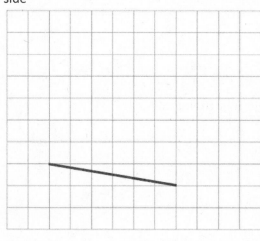

d. A quadrilateral that has at least one pair of perpendicular sides and at least one pair of parallel sides.

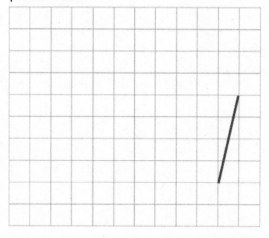

EUREKA
MATH™

**Lesson 16:** Reason about attributes to construct quadrilaterals on square or triangular grid paper.

121

©2018 Great Minds®. eureka-math.org

2. On the grid paper, draw at least one quadrilateral to fit the description. Use the given segment as one segment of the quadrilateral. Name the figure you drew using one of the terms below.

| Parallelogram | Trapezoid | Rectangle |
|---|---|---|
| Square | | Rhombus |

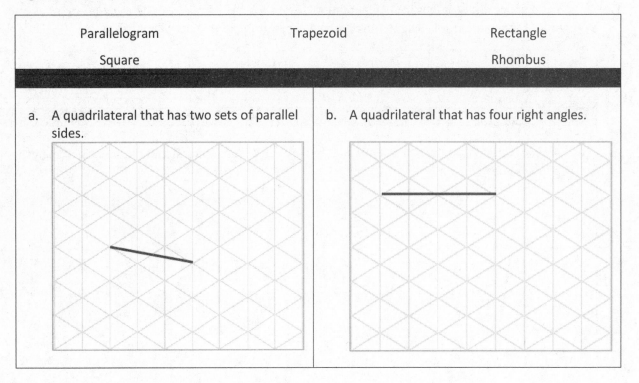

a. A quadrilateral that has two sets of parallel sides.

b. A quadrilateral that has four right angles.

3. Explain the attributes that make a rhombus different from a rectangle.

4. Explain the attribute that makes a square different from a rhombus.

Lesson 16:   Reason about attributes to construct quadrilaterals on square or triangular grid paper.

©2018 Great Minds®. eureka-math.org

EUREKA MATH™

Name _____ Date _____

1. Construct a parallelogram that does not have any right angles on a rectangular grid.

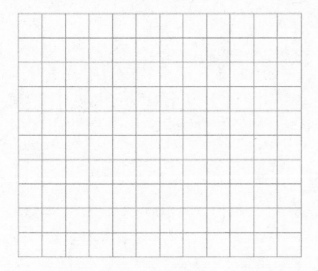

2. Construct a rectangle on a triangular grid.

**EUREKA MATH**

**Lesson 16:** Reason about attributes to construct quadrilaterals on square or triangular grid paper.

©2018 Great Minds®. eureka-math.org

**123**

# Credits

Great Minds® has made every effort to obtain permission for the reprinting of all copyrighted material. If any owner of copyrighted material is not acknowledged herein, please contact Great Minds for proper acknowledgment in all future editions and reprints of this module.